The Open University

Mathematics/Science/Technology
An Inter-Faculty Second Level Course

Mechanics and Applied Calculus

Unit 12 FOURIER ANALYSIS AND NORMAL MODES

Prepared by the Course Team

The Open University Press

The Open University Press, Walton Hall, Bletchley, Bucks.

First published 1972

Designed by the Media Development Group of the Open University.

Printed in Great Britain by
Martin Cadbury Printing Group

SBN 335 01177 2

This text forms part of the correspondence element of an Open University Second Level Course. The complete list of units in the course is given at the end of this text.

For general availability of supporting material referred to in this text, please write to the Director of Marketing, The Open University, Walton Hall, Bletchley, Buckinghamshire.

Further information on Open University courses may be obtained from the Admissions Office, The Open University, P.O. Box 48, Bletchley, Buckinghamshire.

1.1

Contents

Bibliography

W. Kaplan, *Advanced Calculus* (Addison Wesley, 1962).

D. L. Kreider, R. G. Kuller, D. R. Ostberg and F. W. Perkins, *An Introduction to Linear Analysis* (Addison Wesley, 1966).

These books give a fuller mathematical treatment of Fourier Analysis than that given in this text.

Note

References to the Open University Mathematics Foundation Course Units (The Open University Press, 1971) take the form *Unit M100 3, Operations and Morphisms.*

References to the Open University Science Foundation Course Units (The Open University Press, 1971) take the form *Unit S100 29, Quantum Theory.*

Objectives

Our aims in this unit are two-fold. First, we aim to round off our discussions on vibrations by looking at vibrating strings and air columns. This study will involve functions of more than one variable, and a corresponding partial differential equation—the wave equation.

Second, we wish to acquaint you with a method of solving the wave equation, and also to introduce *Fourier series*, which provide a useful representation of a periodic function.

After working through this unit you should be able to:

(i) define the terms:
   Fourier series
   odd and even functions
   harmonic waves
   stationary waves
   nodes and anti-nodes
   normal modes
   fundamental mode
   overtone;

(ii) determine whether a function is odd, even or neither odd nor even;

(iii) determine the coefficients in the Fourier series of simple functions of the type considered in the text;

(iv) derive an expression for the one-dimensional wave equation of
   (*a*) a transverse wave in a stretched string, in terms of the tension in the string and the string's mass per unit length,
   and
   (*b*) a longitudinal sound wave, in terms of the appropriate elastic modulus and the density of the medium;

(v) state the boundary conditions and determine the normal modes of one-dimensional vibrating systems similar to those considered in this text.

Study Sequence

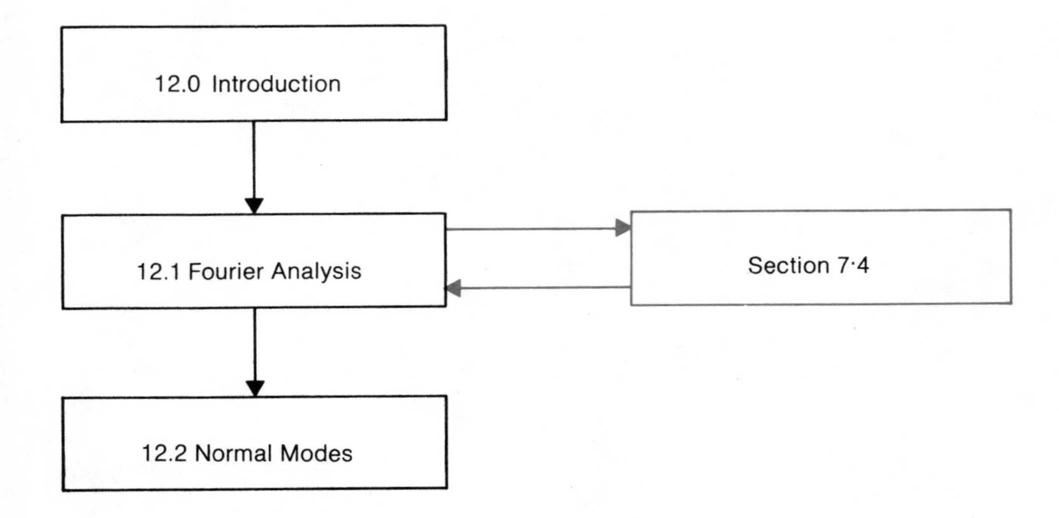

12.0 INTRODUCTION

In *Units 10* and *11*, *Mechanical Vibrations I* and *II*, our equations of motion for forced mechanical vibrations included periodic forcing functions of simple harmonic form:

$$F_0 \cos(\omega t + \delta).$$

However, we are sometimes interested in motion with a forcing function whose form, though periodic, is not necessarily harmonic, that is, in a force, $F(t)$, such that

$$F(t) = F(t + nT),$$

where $n = 0, 1, 2, \ldots$, and T is the period.

The equation of motion we shall consider in this unit still has the form

$$m\ddot{x} + c\dot{x} + kx = F(t),$$

where m, c and k are constants. From *Unit 11*, we know how to deal with such an equation when $F(t)$ is harmonic, but we have not yet considered how to solve it in the more general case when $F(t)$ is periodic but not harmonic. In the following sections, you will see how a special series, known as a *Fourier series*, enables us to express the image under a periodic function as an infinite series of harmonic terms. Using this result, the above equation of motion, with a non-harmonic forcing function, can be interpreted using the results of *Unit 11*. This forms the material of the first part of this text.

In the second part, we take another look at the manner in which physical systems vibrate. For simplicity, we consider the case of vibrations in one dimension—a violin string, for example. We find that the motion can be described in terms of stationary waves. The system vibrates at certain specific frequencies and not at intermediate frequencies—in other words, we can regard the system as having a set of "natural" frequencies. We then point out that *any* vibratory motion of the string can be regarded as the sum of contributions from the modes corresponding to these "natural" frequencies.

This brings us to a further application of the Fourier series method, namely the analysis of the vibratory response of a system. Indeed, this mathematical technique can be employed in many diverse physical problems, not just those associated with mechanical vibrations.

12.1 FOURIER ANALYSIS

12.1.0 Introduction

Previously in this course (for example, in *Unit 4*, section 4.2.1), when we wished to express a general function $f(x)$ in a simpler form we used the linear approximation,

$$f(x) \simeq a_0 + a_1 x,$$

where a_0 and a_1 are determined by evaluating $f(0)$ and $f'(0)$. This is a particular case of the more general Taylor expansion,

$$f(x) \simeq a_0 + a_1 x + a_2 x^2 + \cdots + a_n x^n,$$

where the $a_0, \ldots, a_n$ are determined by evaluating the appropriate derivatives at $x = 0$. Thus the Taylor expansion expresses $f(x)$ approximately as a linear combination of the "simpler" functions $1, x, x^2, x^3, \ldots, x^n$.

We have now reached a similar situation with vibrations. Forces generated by rotating machinery, for example, may be periodic but not necessarily harmonic in form. So we wish to solve the differential equation

$$m\ddot{x} + c\dot{x} + kx = G(t)$$

when $G(t)$ is periodic (i.e. $G(t) = G(t + 2n\pi)$, where n is an integer), but is not harmonic. Can we solve this problem by considering a simpler problem?

By way of answer, we note first that we are able to solve the equation when $G(t)$ is harmonic:

$$m\ddot{x} + c\dot{x} + kx = a \sin(\omega t + \delta) \qquad (\text{or } a \cos(\omega t + \delta)).$$

Moreover, because the differential equation is linear and the sum of two or more solutions is itself a solution, we can immediately conclude that we know how to solve the equation when $G(t)$ is not harmonic but consists of a sum of harmonic terms. The question then becomes: can all periodic functions $G(t)$ be expressed as the sum of harmonic terms? The answer, subject to a few conditions that are not very restrictive, is yes.

Our problem now is to determine the coefficients in such an expression; that is, to evaluate $A_0, A_1, \ldots$, and $B_1, B_2, \ldots$, in the expansion

$$G(t) = \tfrac{1}{2}A_0 + A_1 \cos t + A_2 \cos 2t + \cdots + B_1 \sin t + B_2 \sin 2t + \cdots \qquad (1)$$

(The term $\tfrac{1}{2}A_0$ will be explained later.)

The method of obtaining this expansion is called *Fourier Analysis**.

* For a fuller mathematical treatment of Fourier series than that given in this text, consult Chapter 7 of Kaplan, *Advanced Calculus*. or Kreider *et al.*, *An Introduction to Linear Analysis*. (See Bibliography.)

12.1.1 How to Determine the Coefficients A_n and B_n

Instead of differentiating to determine the coefficients in the expansion

$$G(t) = \tfrac{1}{2}A_0 + A_1 \cos t + A_2 \cos 2t + \cdots + B_1 \sin t + B_2 \sin 2t + \cdots, \quad (1)$$

as with Taylor's expansion, we now form special integrals to remove all the coefficients from the right-hand side except the one which we wish to determine. To do this, we need to use the following definite integrals (which you may care to check, if you have the time, using the appropriate trigonometric identities):

$$\int_{-\pi}^{\pi} \cos nt\, dt = 0 = \int_{-\pi}^{\pi} \sin nt\, dt \qquad (n \in Z, n \neq 0) \quad (2a)$$

$$\int_{-\pi}^{\pi} \cos nt \cos rt\, dt = 0 = \int_{-\pi}^{\pi} \sin nt \sin rt\, dt \qquad (n \in Z, r \in Z, n \neq r) \quad (2b)$$

$$\int_{-\pi}^{\pi} \cos nt \sin rt\, dt = 0 \qquad (n \in Z, r \in Z) \quad (2c)$$

$$\int_{-\pi}^{\pi} \cos^2 nt\, dt = \pi = \int_{-\pi}^{\pi} \sin^2 nt\, dt. \qquad (n \in Z, n \neq 0, r \in Z) \quad (2d)$$

As an example of the general method, we shall calculate A_2. Multiplying Equation (1) by $\cos 2t$, we obtain

$$\begin{aligned} G(t) \cos 2t = \tfrac{1}{2}A_0 \cos 2t &+ A_1 \cos t \cos 2t + A_2 \cos^2 2t \\ &+ A_3 \cos 3t \cos 2t + \cdots \\ &+ B_1 \sin t \cos 2t + B_2 \sin 2t \cos 2t \\ &+ B_3 \sin 3t \cos 2t + \cdots \end{aligned}$$

Now observe what happens to the right-hand side when we integrate from $-\pi$ to π:

$$\begin{aligned} \int_{-\pi}^{\pi} G(t) \cos 2t\, dt = \tfrac{1}{2}A_0 &\int_{-\pi}^{\pi} \cos 2t\, dt + A_1 \int_{-\pi}^{\pi} \cos t \cos 2t\, dt \\ &+ A_2 \int_{-\pi}^{\pi} \cos^2 2t\, dt + A_3 \int_{-\pi}^{\pi} \cos 3t \cos 2t\, dt + \cdots \\ &+ B_1 \int_{-\pi}^{\pi} \sin t \cos 2t\, dt + B_2 \int_{-\pi}^{\pi} \sin 2t \cos 2t\, dt \\ &+ B_3 \int_{-\pi}^{\pi} \sin 3t \cos 2t\, dt + \cdots \end{aligned}$$

We know from Equations (2a)–(2d) that the only non-zero term on the right-hand side is

$$A_2 \int_{-\pi}^{\pi} \cos^2 2t\, dt,$$

and that the value of this term is in fact

$$A_2 \pi.$$

Thus the equation becomes

$$\int_{-\pi}^{\pi} G(t) \cos 2t\, dt = A_2 \pi,$$

and hence

$$A_2 = \frac{1}{\pi}\int_{-\pi}^{\pi} G(t)\cos 2t\, dt.$$

Using a similar method, we find that

$$A_0 = \frac{1}{\pi}\int_{-\pi}^{\pi} G(t)\, dt, \tag{3a}$$

and, in general,

$$A_n = \frac{1}{\pi}\int_{-\pi}^{\pi} G(t)\cos nt\, dt \qquad n \geqslant 1, \tag{3b}$$

$$B_n = \frac{1}{\pi}\int_{-\pi}^{\pi} G(t)\sin nt\, dt \qquad n \geqslant 1. \tag{3c}$$

Now you can see that the constant term in the expansion is written as $\frac{1}{2}A_0$ so that the formulas for A_0 and $A_n (n \geqslant 1)$ are similar. Notice also that, by definition, $\frac{1}{2}A_0$ is the mean value of $G(t)$ over the interval $[-\pi, \pi]$.

Equation (1) may be expressed more concisely as

$$G(t) = \tfrac{1}{2}A_0 + \sum_{n=1}^{\infty} (A_n \cos nt + B_n \sin nt). \tag{4}$$

The right-hand side is known as the Fourier series for $G(t)$, and A_n, B_n are the Fourier coefficients. It can be proved that a given function $G(t)$, which is periodic with period 2π and which satisfies certain properties of continuity in the interval $[-\pi, \pi[$, has a *unique* representation as a Fourier series, but this proof is beyond the scope of this course. A proof is given in Kaplan, *Advanced Calculus*, Chapter 7. (See Bibliography.)

We shall now work through an example, to show how a function can be expanded in a Fourier series.

Example

Consider the periodic function G defined by

$$G(t) = G(t + 2\pi) \text{ for all } t, \text{ and } G(t) = t \text{ for } t \in [-\pi, \pi[.$$

We shall find the Fourier series expansion of this function. (Note that we have excluded one of the end-points of the interval because otherwise $G(\pi) \neq G(-\pi)$.)

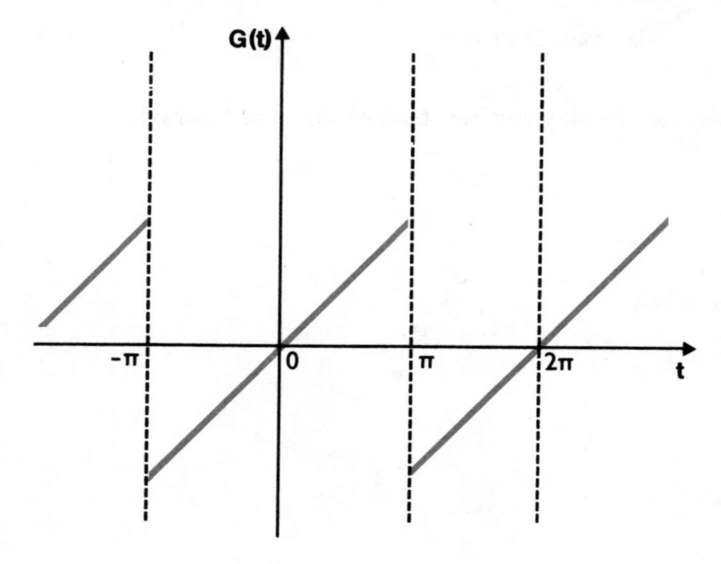

From Equations (3), we have

$$A_0 = \frac{1}{\pi}\int_{-\pi}^{\pi} t\,dt = \frac{1}{\pi}\left[\frac{t^2}{2}\right]_{-\pi}^{\pi} = 0.$$

Integrating by parts, we find

$$\begin{aligned}
A_n &= \frac{1}{\pi}\int_{-\pi}^{\pi} t\cos nt\,dt \\
&= \frac{1}{\pi n}\Big[t\sin nt\Big]_{-\pi}^{\pi} - \frac{1}{\pi}\int_{-\pi}^{\pi}\frac{\sin nt}{n}\,dt \\
&= 0 \\
B_n &= \frac{1}{\pi}\int_{-\pi}^{\pi} t\sin nt\,dt \\
&= \frac{1}{\pi n}\Big[-t\cos nt\Big]_{-\pi}^{\pi} + \frac{1}{\pi}\int_{-\pi}^{\pi}\frac{\cos nt}{n}\,dt \\
&= \frac{1}{\pi n}\Big[-t\cos nt\Big]_{-\pi}^{\pi} = -\frac{2\cos n\pi}{n}.
\end{aligned}$$

Now $\cos n\pi = (-1)^n$, so we have

$$A_n = 0 \qquad B_n = (-1)^{n+1}\frac{2}{n}.$$

Equation (4) now becomes

$$t = 2\left(\sin t - \frac{\sin 2t}{2} + \frac{\sin 3t}{3} - \frac{\sin 4t}{4} + \cdots\right) \qquad t \in [-\pi, \pi[.$$

Note that $G(t)$ is not continuous; at $t = \pm\pi$, $\pm 3\pi$, etc. there are "jump" discontinuities. What happens to the Fourier series at such points of discontinuity? The Fourier series behaves in a remarkably sensible way. It converges to the *average* of the right and left-hand limits. If you require a proof of this, see Kaplan, *Advanced Calculus*, sections 7–8, 7–9. (See Bibliography.)

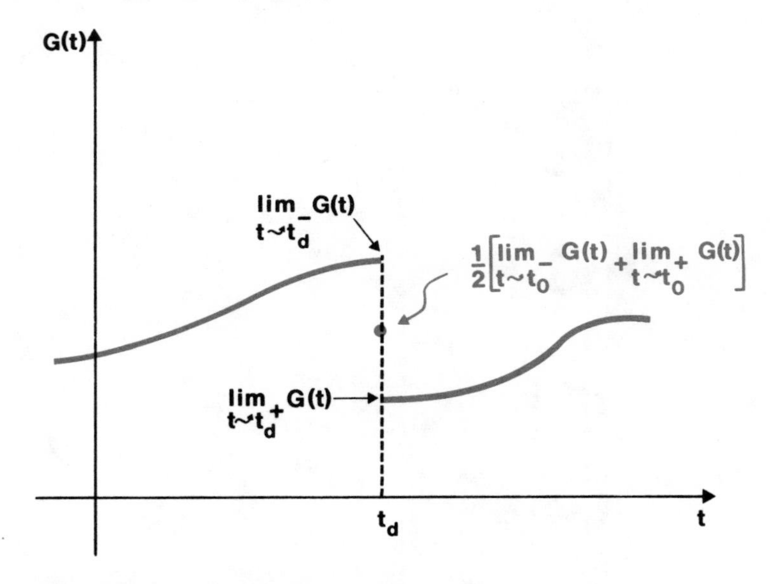

Thus, for a discontinuity at t_d the series converges to

$$\frac{1}{2}\left(\lim_{t\to t_d^-} G(t) + \lim_{t\to t_d^+} G(t)\right)$$

at t_d, where $\lim_{t \to t_d^-} G(t)$ and $\lim_{t \to t_d^+} G(t)$ denote the limits of $G(t)$ as t approaches t_d "from the left" and "from the right" respectively. In the case of the example we have just considered, at $t = \pm\pi$, $\pm 3\pi$, etc. the series converges to the value

$$\frac{1}{2}\left(\lim_{t \to \pi^-} t + \lim_{t \to \pi^+} (t - 2\pi)\right)$$

$$= \tfrac{1}{2}(\pi - \pi) = 0.$$

SAQ 1

Expand

$$G(t) = \begin{cases} 0 & -\pi \leqslant t \leqslant 0 \\ \sin t & 0 < t \leqslant \pi \end{cases}$$

in a Fourier series in the interval $[-\pi, \pi]$.

(Solution is given on p. 30.)

SAQ 2

Expand $G(t) = t^2$ in a Fourier series in the interval $[-\pi, \pi]$ and deduce that

$$\sum_{n=1}^{\infty} \frac{1}{n^2} = \frac{\pi^2}{6}.$$

(Solution is given on p. 30.)

12.1.2 Odd and Even Functions and Their Fourier Series in $[-\pi, \pi[$

We define an odd function to be a function f such that

$$f(t) = -f(-t) \qquad (t \in R)$$

and an even function to be a function f such that

$$f(t) = f(-t) \qquad (t \in R)$$

SAQ 3

Indicate whether the following functions are odd, even, or neither. In each case, $t \in R$.

(i) $f(t) = t$	(ii) $f(t) = t^2$
(iii) $f(t) = e^t$	(iv) $f(t) = t^2 - t$
(v) $f(t) = \sin 3t$	(vi) $f(t) = \sin 4t$
(vii) $f(t) = 1$	(viii) $f(t) = \cos t + \cos 2t$
(ix) $f(t) = t \sin t$	(x) $f(t) = \begin{cases} t^2 \sin 2t & t \geqslant 0 \\ -t^2 \sin 2t & t < 0. \end{cases}$

(Solution is given on p. 31.)

The important feature to note from the last SAQ is that all *sine* functions are *odd* and all *cosine* functions and *constant* functions are *even*. One would therefore suspect that the Fourier series for an *odd* function would contain only *sine* terms, whereas the Fourier series for an *even* function would contain only the *constant* term and *cosine* terms. We ask you to confirm the latter statement in the next SAQ.

SAQ 4

Show that, if $f(t)$ is even and $f(t + 2\pi) = f(t)$ for all $t \in R$, then the coefficient of $\sin nt$ in the Fourier series expansion of f is zero.

HINT: Split the interval of integration, $]-\pi, \pi[$, into two equal intervals.

(Solution is given on p. 31.)

SAQ 5

Find the Fourier series for

$$G(t) = \begin{cases} -1 & -\pi \leqslant t \leqslant 0 \\ 1 & 0 < t < \pi \end{cases}$$

and sketch the periodic function given by

$$G(t + 2\pi) = G(t) \qquad (t \in R).$$

(Solution is given on p. 32.)

SAQ 6

Find the Fourier series for

$$G(t) = \begin{cases} -\sin t & -\pi \leqslant t \leqslant 0 \\ \sin t & 0 < t \leqslant \pi \end{cases}$$

and sketch the periodic function given by

$$G(t + 2\pi) = G(t) \qquad (t \in R).$$

(Solution is given on p. 32.)

12.1.3 "Read the Book" Section

We now ask you to read a part of the set book. In it the Fourier series we have been discussing are dealt with very briefly.

Read part of section S7.4, pp. S186–S188.

SAQ 7

Exercise 32, p. S197.

(Solution is given on p. 33.)

12.1.4 Summary

In this section we have concentrated our attention on Fourier series, primarily to give you an understanding of what they are and to enable you to use them if you need to do so. If you wish to look at Fourier series in greater detail, you should consult Chapter 7 of Kaplan, *Advanced Calculus*. If you wish to see how the topic is developed from a strictly mathematical viewpoint, you could refer to Chapter 9 of D. L. Kreider *et al.*, *An Introduction to Linear Analysis*; this book deals with the convergence problems which we have ignored in our approach. (See Bibliography.)

In section 12.1 we have found that, generally speaking, a periodic function $G(t)$ may be conveniently represented by a series of sines and cosines, called a *Fourier series:*

$$G(t) = \tfrac{1}{2}A_0 + \sum_{n=1}^{\infty} (A_n \cos nt + B_n \sin nt),$$

where the coefficients are given by

$$A_0 = \frac{1}{\pi}\int_{-\pi}^{\pi} G(t)\,dt,$$

$$A_n = \frac{1}{\pi}\int_{-\pi}^{\pi} G(t)\cos nt\,dt \qquad n \geqslant 1,$$

$$B_n = \frac{1}{\pi}\int_{-\pi}^{\pi} G(t)\sin nt\,dt \qquad n \geqslant 1.$$

We also defined odd functions ($G(t) = -G(-t)$) and even functions ($G(t) = G(-t)$), and showed that their Fourier series contain sine terms only, and constant and cosine terms only, respectively.

The subject has been presented in the specific context of representing a time-dependent periodic force. But you will see in the next section that there is no reason why we should restrict the application of the Fourier series method to functions of time.

12.2 NORMAL MODES

12.2.0 Introduction

In this section we round off our discussion of vibrations. We again take a look at the way in which a system vibrates. So far, we have considered a simple harmonic oscillator—for example, a particle on the end of a spring, or a simple pendulum consisting of a particle swinging on the end of an inextensible string. The motion of the particle is characterized by the natural frequency of the system. When damping is incorporated, the period of the oscillation is altered, but the system still has only one single characteristic frequency for the unforced vibration. Note that in these systems we are interested in the motion of only one particle, and the spring (or string) is merely considered as the means of exerting a force on the particle.

Now we are interested in *extended* systems, that is, systems containing many particles. Suppose we wish to investigate the vibrations of *all* these particles under the influence of the forces exerted by the particles on each other. This is a much more interesting situation. In particular, we find that the system as a whole can have *many* "natural" frequencies. Depending on how the motion is started, that is, what values of displacement and velocity are initially given to the particles in each part of the system, we get different vibrational motions corresponding to any one of these characteristic natural frequencies. In the general case, the system vibrates in a manner that can be regarded as a *combination* of these individual modes. The analysis of the motion in terms of these component modes can often be achieved through the use of the Fourier analysis technique introduced in the previous section. As an example, we shall show that the complicated motion of a violin string can be analysed in this way.

12.2.1 Wave Motion in One Dimension

We consider an extended system, composed of many particles. Each part of the system is capable of exerting a force on its neighbouring parts. For simplicity we shall take the one-dimensional case of a stretched string.

If part of the string is displaced laterally from its equilibrium position, it exerts forces on the neighbouring segments of the string on either side. As a result, these segments are no longer in equilibrium. They also move laterally, and in so doing act in turn upon their neighbours—and so on. In effect, a disturbance moves progressively down the string in both directions. Such behaviour is a characteristic of *wave motion.*

These considerations give us a clue as to how we might tackle the problem of the vibrating string. At the moment when the string is released, its various parts are displaced from their equilibrium positions. The subsequent motion arising from such displacements can be regarded as waves propagating along the string. So this leads us to consider wave motion.

Consider a string, of mass ρ per unit length, which, in equilibrium, is stretched and lies along the x-axis. Let the vibrations displace it transversely by an amount y at distance x, and let the tension be T throughout the string. For small displacements, we may assume that T is constant.

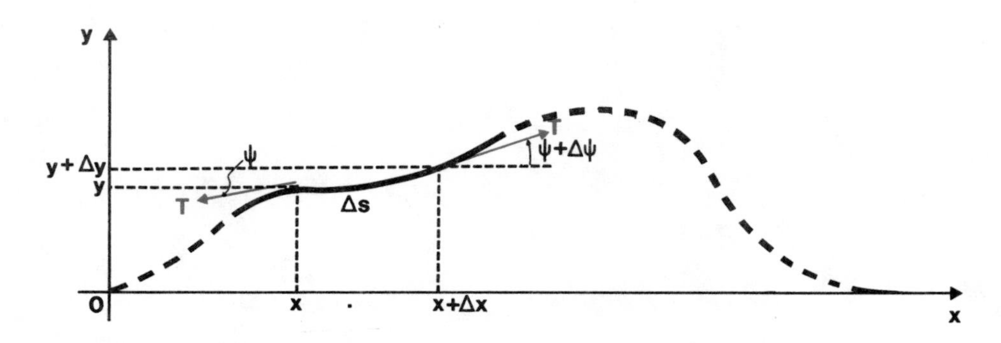

Consider a small element Δs of the string at some instant. Assume that the tangent to the string at (x, y) makes an angle ψ with the x-axis, and that at $(x + \Delta x,\ y + \Delta y)$ it makes an angle $\psi + \Delta\psi$. Using Newton's second law, we find that the equation of motion of the element in the y-direction (ignoring gravity) is

$$\underbrace{T \sin(\psi + \Delta\psi) - T \sin\psi}_{\text{resultant force in } y\text{-direction}} = \underbrace{\frac{\partial^2 y}{\partial t^2}}_{\substack{\text{acceleration} \\ \text{in } y\text{-direction}}} \times \underbrace{\rho\, \Delta s}_{\text{mass}}.$$

Note that we need to use *partial* derivatives, since the variable y is dependent on both x and t.

Using the relation

$$\sin(\psi + \Delta\psi) = \sin\psi \cos\Delta\psi + \cos\psi \sin\Delta\psi,$$

and neglecting squares of small quantities (i.e. using the linear Taylor approximation for $\cos\Delta\psi$ and $\sin\Delta\psi$), we find

$$\begin{aligned}\cos\Delta\psi &= 1 - \tfrac{1}{2}(\Delta\psi)^2 + \cdots \\ &\simeq 1\end{aligned}$$

and

$$\begin{aligned}\sin\Delta\psi &= \Delta\psi - \tfrac{1}{6}(\Delta\psi)^3 + \cdots \\ &\simeq \Delta\psi.\end{aligned}$$

The equation of motion becomes approximately

$$T \cos\psi\, \Delta\psi = \rho\, \Delta s \frac{\partial^2 y}{\partial t^2}.$$

Now, from the figure, we have

$$\Delta s \simeq \Delta x \sec\psi$$

so

$$T \cos^2\psi \frac{\Delta\psi}{\Delta x} = \rho \frac{\partial^2 y}{\partial t^2}.$$

In the limit as Δx approaches zero, this becomes

$$T \cos^2\psi \frac{\partial\psi}{\partial x} = \rho \frac{\partial^2 y}{\partial t^2}. \qquad (5)$$

At any given instant, we know that

$$\tan\psi = \frac{\partial y}{\partial x},$$

so

$$\frac{\partial(\tan\psi)}{\partial x} = \frac{\partial^2 y}{\partial x^2},$$

and

$$\sec^2\psi \frac{\partial\psi}{\partial x} = \frac{\partial^2 y}{\partial x^2}.$$

Hence

$$\frac{\partial \psi}{\partial x} = \cos^2 \psi \frac{\partial^2 y}{\partial x^2}.$$

Thus, on substituting in Equation (5), we obtain

$$T \cos^4 \psi \frac{\partial^2 y}{\partial x^2} = \rho \frac{\partial^2 y}{\partial t^2}.$$

Finally,

$$\cos^4 \psi = \frac{1}{\sec^4 \psi} = \frac{1}{(1 + \tan^2 \psi)^2} = \frac{1}{\left[1 + \left(\frac{\partial y}{\partial x}\right)^2\right]^2}.$$

If the displacements are very small compared with the length of the string (and this is true in modelling cases such as the vibrating violin string), then $(\partial y/\partial x)^2 \ll 1$, and the equation of wave motion is approximately

$$\frac{\partial^2 y}{\partial x^2} = \frac{\rho}{T} \frac{\partial^2 y}{\partial t^2}. \tag{6}$$

(Note that the assumption made in the derivation of this equation corresponds to the assumption that the angle ψ remains small.) The equation is generally written in the form

$$\frac{\partial^2 y}{\partial x^2} = \frac{1}{c^2} \frac{\partial^2 y}{\partial t^2}, \tag{7}$$

where $c^2 = T/\rho$. Equation (7) is the wave equation in one dimension. It describes the motion of a stretched string rather well if the displacement y is not large. Notice that it is a *linear* equation. Thus, if y_1 and y_2 are any two solutions, $A_1 y_1 + A_2 y_2$ is also a solution, A_1 and A_2 being arbitrary constants.

12.2.2 The General Solution of the Wave Equation

The most general solution of Equation (7) is due to D'Alembert, who discovered the solution of the form

$$y = f(x - ct) + g(x + ct), \tag{8}$$

where f and g are "well-behaved" functions. We now derive this equation.

We change from variables x, t in the original differential equation to variables u, v, where

$$u = x - ct, \qquad v = x + ct.$$

The chain rule for functions of two variables gives

$$\frac{\partial y}{\partial x} = \frac{\partial y}{\partial u}\frac{\partial u}{\partial x} + \frac{\partial y}{\partial v}\frac{\partial v}{\partial x} = \frac{\partial y}{\partial u} + \frac{\partial y}{\partial v},$$

$$\frac{\partial^2 y}{\partial x^2} = \frac{\partial}{\partial x}\frac{\partial y}{\partial x} = \frac{\partial}{\partial u}\left(\frac{\partial y}{\partial u} + \frac{\partial y}{\partial v}\right) + \frac{\partial}{\partial v}\left(\frac{\partial y}{\partial u} + \frac{\partial y}{\partial v}\right)$$

$$= \frac{\partial^2 y}{\partial u^2} + 2\frac{\partial^2 y}{\partial u\,\partial v} + \frac{\partial^2 y}{\partial v^2}.$$

You might like to show similarly that

$$\frac{\partial^2 y}{\partial t^2} = c^2\frac{\partial^2 y}{\partial u^2} - 2c^2\frac{\partial^2 y}{\partial u\,\partial v} + c^2\frac{\partial^2 y}{\partial v^2}.$$

Substitution for $\frac{\partial^2 y}{\partial x^2}$ and $\frac{\partial^2 y}{\partial t^2}$ in the wave equation, followed by simplification, gives

$$\frac{\partial^2 y}{\partial u\,\partial v} = 0.$$

We can integrate this equation to give either

$$\frac{\partial y}{\partial u} = F(u)$$

or

$$\frac{\partial y}{\partial v} = G(v),$$

where F and G are any "well-behaved" functions of u and v respectively,

Integration a second time gives

$$y = f(u) + g(v),$$

where

$$f(u) = \int F(u)\,du \qquad \text{and} \qquad g(v) = \int G(v)\,dv.$$

Substitution for u and v in terms of x and t produces D'Alembert's solution—Equation (8).

What sort of solutions do these functions represent? Let us first consider $y = f(x - ct)$. Suppose that, at $t = 0$, the disturbance is given by

$$y = f(x).$$

The graph of the disturbance y as a function of x is called the *profile* of the wave. At a later time, $t > 0$, the profile is represented by

$$y = f(x - ct).$$

But if we measure from a new origin at $x = ct$ that is, if we put $x = X + ct$, then the wave profile referred to this new origin is $y = f(X)$.

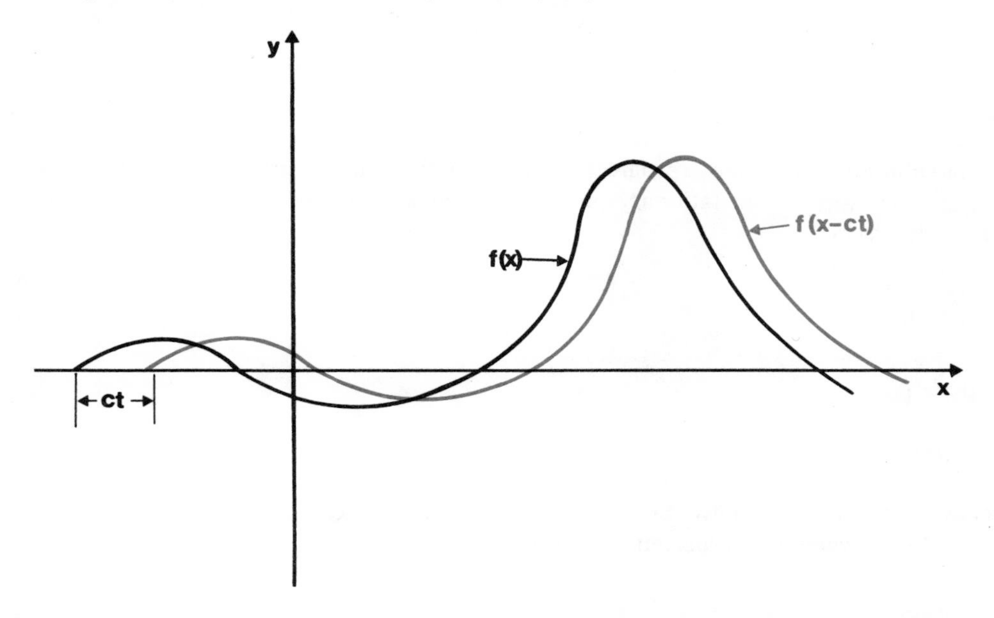

Thus the wave has moved a distance ct in the positive x-direction *without changing its shape.* As the wave has moved the distance ct in time t, we see that the significance of the constant c is that it represents the *velocity* of the wave. The equation $y = f(x - ct)$ is the most general expression for a travelling wave moving with velocity c in the positive x-direction.

What does $y = g(x + ct)$ represent? Well, in order to change from a variable of the form $(x - ct)$ to one of the form $(x + ct)$, we merely replace c by $-c$; in other words, we reverse the direction of the velocity. Thus, $y = g(x + ct)$ must represent a travelling wave, with profile given by $y = g(x)$ at $t = 0$, moving in the *negative* x-direction.

In this way we see that the solution

$$y = f(x - ct) + g(x + ct)$$

is the most general solution in that it allows for waves to be moving in either direction.

The simplest examples of waves are sine and cosine waves, having, for example, the form

$$y = A \sin p(x + ct)$$

or

$$y = B \cos p(x - ct).$$

These are called harmonic waves. The maximum value of the disturbance, A or B, is called the amplitude. The constant p is given by $p = 2\pi/\lambda$, where λ is the wavelength.

12.2.3 Stationary Harmonic Waves

Since the wave equation is a linear differential equation, we know that we can add any two solutions to form another solution. If we add together two cosine waves with the same amplitude and period, we get, for example,

$$y = A \cos p(x - ct) + A \cos p(x + ct)$$

i.e.

$$y = 2A \cos px \cos pct.$$

Because the profile of this solution does *not* move progressively along in the x-direction, this wave is called a stationary wave. The stationary property of the wave is easily appreciated by looking at the points

$$x = \pm\frac{\pi}{2p}, \pm\frac{3\pi}{2p}, \pm\frac{5\pi}{2p}, \ldots,$$

where $\cos px = 0$. At these points,

$$y = 0 \textit{ for all values of } t,$$

and the particles at these positions do not move. Such points, where the displacement is zero, are called nodes. Points where the displacement is a maximum,

$$x = 0, \pm\frac{\pi}{p}, \pm\frac{2\pi}{p}, \ldots,$$

are called anti-nodes. If we look at the motion at any point x, we see an oscillation of fixed amplitude ($2A \cos px$) which varies sinusoidally with time.

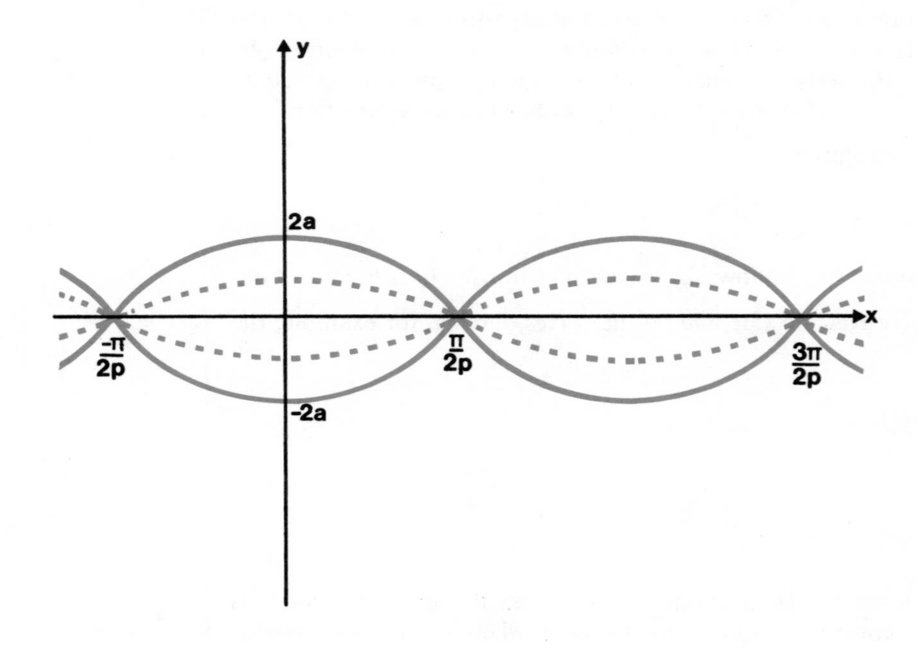

12.2.4 Normal Modes of a Stretched String

You may be wondering what sort of practical situation requires the addition of two solutions of equal amplitude and period. In fact this situation arises very often, particularly with stringed musical instruments.

Suppose a string is firmly clamped at one end. A wave disturbance arriving at this point is reflected. If no energy loss is incurred, then the amplitude of the reflected wave is the same as that of the incident wave.

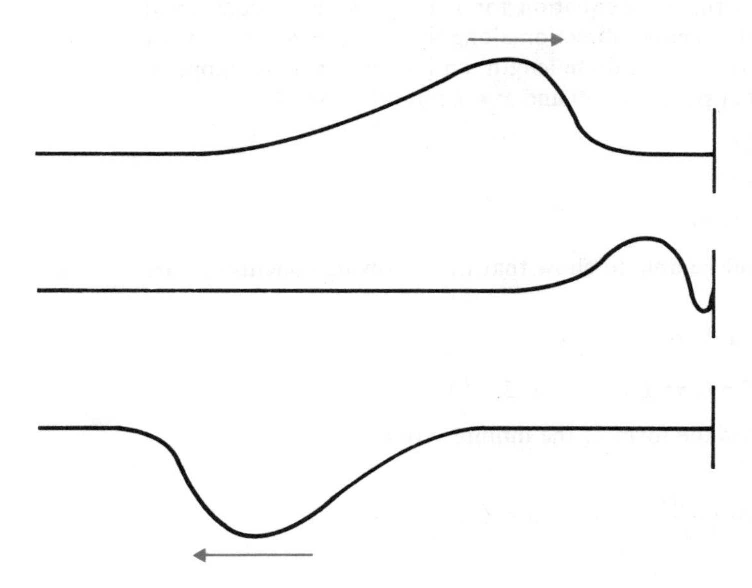

(The displacement has opposite sign because the wave is reflected upside-down.) Moreover, if the incident wave is periodic, the reflected wave will also be periodic and have the same frequency. Thus we can get a situation where part of a periodic wave is approaching the fixed end of the string, while a more advanced part of the wave train is at the same time moving back along the same string *after* reflection. The motion of the clamped string is therefore described by the sum of two solutions, one representing the incident part of the wave train and the other the reflected part, both having the same amplitude and frequency. If the other end of the string is also fixed, then we get repeated reflections. (There must, of course, be *some* yielding of the support of a violin string—otherwise none of the energy would be transmitted to the body of the instrument. But, even so, this yielding is so small that for most purposes the end can be regarded as fixed.)

We can extend the idea of adding together harmonic solutions in order to get additional solutions. Often, a solution is given by a finite combination of sine and cosine terms,

$$y = \sum_p (A_p \cos px + B_p \sin px)(\cos pct + C_p \sin pct),$$

where $p = 2\pi/\lambda$, and the wavelength λ can take any positive value, but takes only a finite number of values. ($\sum_p$ denotes summation over these values). Note that there is no need to include a coefficient for cos pct; such a coefficient would be redundant, for we could take it out as a factor of the second bracket and incorporate it in the coefficients A_p and B_p.

We can extend this further, by removing the restriction that p takes only a finite number of values. Then, provided that the infinite series $\sum_p A_p$, $\sum_p B_p$ and $\sum_p C_p$ all converge, another solution is given by the infinite series

$$y = \sum_p (A_p \cos px + B_p \sin px)(\cos pct + C_p \sin pct). \qquad (9)$$

This form of solution enables us to predict the behaviour of disturbances with almost any initial shape.

An alternative useful form of Equation (9), using an appropriate trigonometric identity, is

$$y = \sum_p (A_p \cos px + B_p \sin px) \cos (pct + \varepsilon_p). \qquad (10)$$

(The A_p's and B_p's are now different, and we have introduced the phase angle ε_p.)

In writing the above solution to the wave equation for a string, we have considered waves that can extend indefinitely in either direction along the string. When this is not the case and the waves are restricted to a finite length, an interesting thing happens. Suppose that the string is *fixed* at points $x = 0$ and $x = L$; in other words,

(i) $y = 0$ at $x = 0$ for all t;
(ii) $y = 0$ at $x = L$ for all t.

These are called *boundary conditions.*

Using Equation (10), you should be able to show that the following conclusions are immediately forthcoming:

condition (i) is satisfied only if $A_p = 0$ for all p;

condition (ii) is satisfied only if $p = n\pi/L$ $(n = 1, 2, 3, \ldots)$.

Thus in this case the solution has the form of the infinite series

$$y = \sum_{n=1}^{\infty} B_n \sin\left(\frac{n\pi x}{L}\right) \cos\left(\frac{n\pi ct}{L} + \varepsilon_n\right) \qquad n \in Z^+,$$

(where we have substituted for p).

Each of the terms of this series (in which n takes some positive integral value) is called a normal mode. Each normal mode is characterized by its own angular frequency, $n\pi c/L$, or alternatively, by its own frequency, $(n\pi c/L)/2\pi = nc/2L$. These are the "natural" frequencies of the particular system we are considering, namely a string fixed at both ends. The lowest frequency, called the *fundamental*, occurs when $n = 1$, and has a value $c/2L$. Remembering that $p = 2\pi/\lambda$, we see that the requirement $p = n\pi/L$ yields $L = \lambda/2$ for the fundamental mode of vibration. The next mode occurs for $n = 2$, whereupon the frequency is doubled (c/L), and $L = \lambda$. The next mode has $n = 3$, a frequency of $3c/2L$ and $L = 3\lambda/2$. These modes are sketched in the following diagram. The allowed frequencies which are greater than the fundamental are called *overtones.*

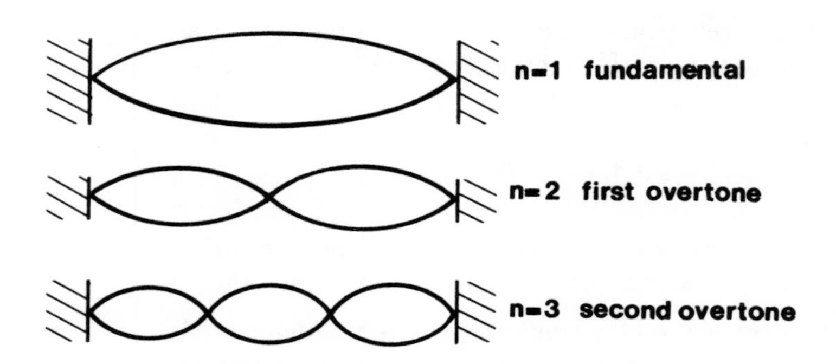

The existence of these normal modes arises from the nature of the boundary conditions. The system can *only* vibrate at these allowed frequencies. It follows therefore that it must be possible to express any general form of vibration of the string in terms of these normal modes. *The particular combination of modes is governed by the initial conditions.* Initial conditions specify the displacement and velocity of each part of the system at time $t = 0$. Let us illustrate how this comes about with the help of an example.

Suppose a stretched string, fixed at points $x = 0$ and $x = L$, is pulled into the shape shown in the following figure, and then released from rest at time $t = 0$.

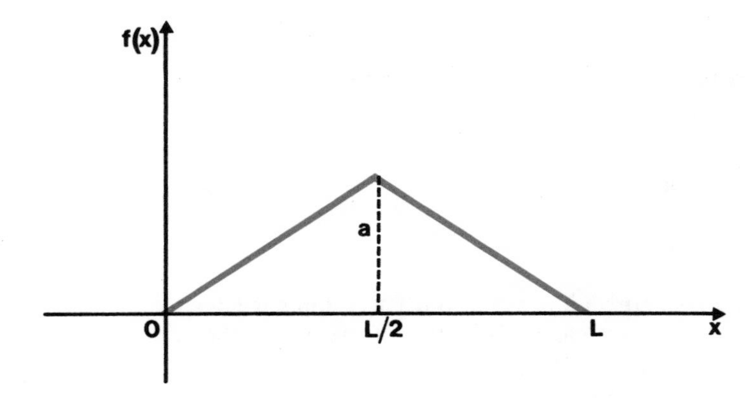

We now have two further conditions to add to the two boundary conditions. These new conditions will enable us to determine the constants B_n and ε_n.

These initial conditions are:

(iii) $y = f(x)$ at $t = 0$, where

$$f(x) = \begin{cases} \dfrac{2ax}{L} & 0 \leqslant x \leqslant \dfrac{L}{2} \\[2ex] \dfrac{2a}{L}(L - x) & \dfrac{L}{2} < x \leqslant L \end{cases}$$

(iv) $\dfrac{\partial y}{\partial t} = 0$ for all x at $t = 0$

(i.e. all parts of the string have zero initial velocity).

We see immediately that condition (iv) can be satisfied only if all the phase angles ε_n are zero.

There remains condition (iii). This is to be used to determine the coefficients B_n in the equation

$$f(x) = \sum_{n=1}^{\infty} B_n \sin\left(\frac{n\pi x}{L}\right) \qquad x \in [0, L],$$

which is a Fourier series representation of the function at $t = 0$.

You may have noticed that the shape of the graph in this example is similar to the shape of the graph in Example 7, p. S187. Indeed, if we set $F_0 = a$ and $2\pi/\omega = L$ in Example 7, the two functions do appear to have the same analytic representation. But there is one essential difference between the two cases, which leads us to choose a different Fourier representation here. In the book the function F is periodic. In the present example it is immaterial how our mathematical representation behaves outside the interval $[0, L]$, for there is nothing to be modelled there. We are interested in finding the magnitudes of the different modes of oscillation, and they will be given by evaluating the B_n's above. The series expansion in Example 7 in S would not give us these values.

In Example 7 on page S187, the function was a function of time; now it is a function of distance, x. This is of no importance; we can once again apply the Fourier analysis method. The solution follows the same lines as before, the integration now being over the interval $[0, L]$ instead of $[0, 2\pi/\omega]$. Thus

$$B_n = \frac{2}{L}\int_0^L f(x) \sin\left(\frac{n\pi x}{L}\right) dx$$

$$= \frac{4a}{L^2}\left[\int_0^{L/2} x \sin\left(\frac{n\pi x}{L}\right) dx + \int_{L/2}^{L} (L - x) \sin\left(\frac{n\pi x}{L}\right) dx\right].$$

Integration by parts of each of these two integrals yields the result

$$B_n = \frac{8a}{n^2\pi^2} \sin\left(\frac{n\pi}{2}\right).$$

Hence

$$B_n = \begin{cases} \dfrac{8a(-1)^{(n-1)/2}}{n^2\pi^2} & (n \text{ odd}) \\ 0 & (n \text{ even}). \end{cases}$$

Thus we see that only modes with n odd are set into vibration. This is quite reasonable. After all, even-numbered modes all have a node at the centre of the string; the odd-numbered ones have an anti-node. We would not expect the former to occur when the string is initially pulled out to its maximum displacement in the centre, as in this example.

Replacing n by $(2m + 1)$, and requiring m to be a positive integer, allows us to write the complete solution in a form that automatically takes into account the absence of even terms,

$$y = \sum_{m=0}^{\infty} \frac{(-1)^m 8a}{(2m+1)^2\pi^2} \sin\left(\frac{(2m+1)\pi x}{L}\right) \cos\left(\frac{(2m+1)\pi ct}{L}\right).$$

SAQ 8

A string lies along the x-axis at time $t = 0$. It is attached, at $x = 0$, to a ring of negligible mass constrained to move along a frictionless wire which lies in a direction normal to the x-axis. Which of the following options is the appropriate mathematical condition for this system?

A $y = 0$ at $x = 0$ for all t

B $\dfrac{\partial y}{\partial x} = 0$ at $x = 0$ for all t

C $y = 0$ at $t = 0$ for all x

D $\dfrac{\partial y}{\partial x} = 0$ at $t = 0$ for all x

E $\dfrac{\partial y}{\partial t} = 0$ at $x = 0$ for all t

(Solution is given on p. 35.)

SAQ 9

The two ends of a uniform vertical string of density ρ and length L, stretched to a tension $T > 0$, are attached to weightless rings which can slide on two horizontal frictionless wires at $z = 0$ and $z = L$. Write down the equation of the string and simplify it as much as possible.

(Solution is given on p. 35.)

SAQ 10

A piano wire of density ρ, stretched to a tension T, is fixed at $x = 0$ and $x = L$. A note is produced by means of a hammer giving a transverse velocity V to the part of the string $x \in [a - \delta, a + \delta]$, the remainder of the string being initially at rest. Find the subsequent displacement as a Fourier series and show that the nth harmonic is missing if an/L is an integer.

(Solution is given on p. 35.)

SAQ 11

Which option completes the following statement correctly? The functions describing the possible shapes of a string stretched between the fixed points at $x = 0$ and $x = L$ ...

A are all odd

B are all even

C can be either odd or even

D cannot be determined as odd or even

(Solution is given on p. 37.)

12.2.5 Normal Modes of Other One-Dimensional Systems

Having looked at the transverse vibrations of a stretched string, we now examine further physical situations modelled by the wave equation; these involve *longitudinal* (i.e. backward and forward) vibrations. In particular, we shall look at the longitudinal vibration of a uniform rod, and of the air in a hollow tube. In both these cases, compressibility of the material is the important factor.

A simple model of longitudinal (sound) wave propagation through a compressible medium is the following. We assume that the medium consists of a large number of identical particles. The sound wave causes the particles to move backwards and forwards in simple harmonic motion about the mean positions of their otherwise random (thermal) motion. The motion is transmitted between the particles by collision.

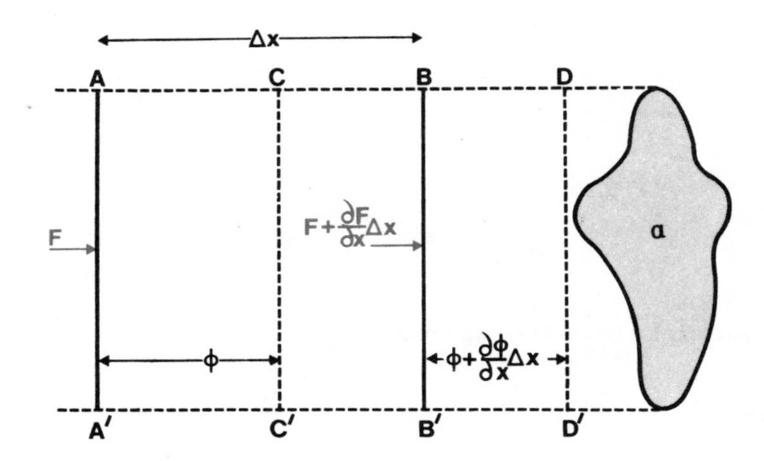

The figure shows an element $ABB'A'$ of a column of air or solid in equilibrium with its surrounding medium (air or solid respectively). The planes AA' and BB' are normal to the direction along which the sound wave passes. The separation of the planes when in equilibrium is $\Delta x(\neq 0)$. Their cross-sectional area is α. When the sound wave arrives, a pressure variation is superimposed on the atmospheric pressure. As a result of this variation, the planes AA' and BB', move to CC' and DD' respectively. The displacement of AA' is represented by the variable ϕ. The distance BD is generally not equal to AC, since the medium is compressible. We therefore write it as

$$BD = \phi + \frac{\partial \phi}{\partial x}\,\Delta x.$$

The force on AA' at any instant is F, and that on BB' is

$$-(F + \Delta F) = -\left(F + \frac{\partial F}{\partial x}\,\Delta x\right).$$

(The negative sign arises because we are considering the force on $ABB'A'$ due to the surrounding medium.) Therefore the resultant force acting on $ABB'A'$ in the direction of increasing ϕ is

$$F - \left(F + \frac{\partial F}{\partial x}\Delta x\right) = -\frac{\partial F}{\partial x}\Delta x.$$

If ρ_0 is the density of the medium (assumed constant to a first order approximation), the equation of motion becomes

$$\rho_0 \alpha \Delta x \frac{\partial^2 \phi}{\partial t^2} = -\frac{\partial F}{\partial x}\Delta x, \qquad (11)$$

where the left-hand side of the equation represents the element's mass multiplied by its acceleration. We now need to relate the force, F, to some property of the medium.

In an elastic medium, the force is usually a linear function of the extension (to a first order approximation). Hooke's law states

force/unit area $= -k$ (extension/unit length),

where k is the *modulus of elasticity* (see *Unit 8*).

Thus

$$\frac{F}{\alpha} = -k\frac{\partial \phi}{\partial x},$$

so

$$F = -k\alpha\frac{\partial \phi}{\partial x}$$

and

$$\frac{\partial F}{\partial x} = -k\alpha\frac{\partial^2 \phi}{\partial x^2}.$$

Substituting in the equation of motion, Equation (11), we obtain

$$\rho_0 \alpha \Delta x \frac{\partial^2 \phi}{\partial t^2} = k\alpha\frac{\partial^2 \phi}{\partial x^2}\Delta x,$$

so

$$\frac{\partial^2 \phi}{\partial x^2} = \frac{\rho_0}{k}\frac{\partial^2 \phi}{\partial t^2}. \qquad (12)$$

Immediately we recognize this to be the characteristic form of the wave equation. By comparison with Equation (7), we deduce that the propagation speed of the disturbance is $(k/\rho_0)^{1/2}$. Obviously this depends upon the elastic modulus, and the value of this will depend upon the nature of the medium. For a uniform rod it will be the familiar *Young's modulus*. For a gas or liquid it will be the so-called *adiabatic bulk modulus*.

Regardless of the particular type of elastic modulus involved however, the essential point to grasp is that, for small displacements, we have ended up with exactly the same form of wave equation as we had for the vibrating stretched string. This means that we should be able to set up stationary waves in, for example, a hollow tube of air. If the ends are closed so that no motion of the air can take place at these extreme positions, we have a situation analogous to that of a stretched string fixed at both ends.

Once again we should obtain nodes—places where there is no motion of the medium. Moreover, the nodes will be spaced in a similar manner to those of the stretched string.

We can represent this situation by plotting the longitudinal displacement, ϕ, as a function of x. (Note that in the physical situation, ϕ lies *along* the x-direction, *not* at right angles to it as indicated on the following graph.) In the diagram we illustrate the mode corresponding to the first overtone. The red arrows show the range of ϕ-values attained in one period at a particular value of x.

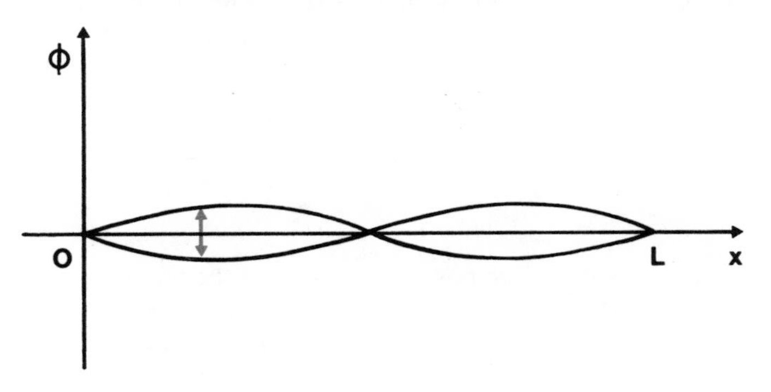

As this is similar to the case of a stretched string, we can immediately carry over the Fourier series method to the analysis of longitudinal vibrations. If either or both of the ends of the tube are not closed, but open (as is the case in musical wind instruments), then it turns out that there is always an anti-node situated at these places. Although this is a different type of boundary condition to the ones encountered so far, it is nevertheless a boundary condition that must be satisfied. Again, there is a restriction which limits the frequencies that can be excited, and gives rise to a set of normal modes.

The ground we have covered in our discussion of stationary waves in one dimension has now given some insight into the behaviour of both string and wind musical instruments. We see now how the frequency of the mode of vibration of a violin string, excited by bowing or plucking, depends upon the length of the string, its tension, and its mass per unit length. Similarly, the note in a wind instrument, excited by blowing on a reed, depends upon the length of the tube.

However, a violin and a trumpet, playing the same note, sound different. Why is this? In fact, neither instrument is producing a single pure frequency. The sound produced in each case contains a mixture of overtones as well as the fundamental. Because the mixture is different in the two cases, each instrument produces its own distinctive sound.

SAQ 12

Two organ pipes of equal length each have an anti-node at the blowing end. Pipe A is open at the far end whereas pipe B is closed.

(i) What is the ratio of the fundamental frequencies of the two pipes?
(ii) Is this the same as the ratio of the frequencies of the first overtones in the two pipes?
(iii) When pipe A sounds its fundamental, what is the wavelength of the sound waves in terms of the length of the pipe?
(iv) If the length of pipe B can be adjusted, would you expect there to be a length such that the sound produced was indistinguishable from that produced by pipe A?

(Solution is given on p. 37.)

12.2.6 Normal Modes in Two- and Three-Dimensional Systems

We have deliberately restricted our discussion to waves in one dimension, but it must already have occurred to you that this could be regarded as merely a beginning. What about waves in two and three dimensions? We end our work on vibrations by mentioning one or two aspects of these more complicated cases.

Consider a rectangular two-dimensional elastic membrane—perhaps a plate glass window fixed at the edges in a frame—which is set vibrating by a sonic boom. If part of the membrane is displaced transversely (in a direction normal to the plane of the membrane), the disturbance propagates outwards like a wave. The wave equation now becomes

$$\frac{\partial^2 \phi}{\partial x^2} + \frac{\partial^2 \phi}{\partial y^2} = \frac{1}{c^2}\frac{\partial^2 \phi}{\partial t^2},$$

where $\phi(x, y, t)$ is the displacement from the equilibrium position at time t at the point (x, y). If the edges of the membrane are fixed, then reflections take place and stationary waves can be established. Instead of nodal points (as we found with the vibrating string), we now have *nodal lines*. Generally, for a rectangular membrane, there are two sets of nodal lines, one set comprising lines parallel to two of the sides, and the other set comprising lines parallel to the other two sides.

The pattern of nodal lines changes as the boundary conditions are altered. The membrane could be circular with the edges either fixed, as in the case of a drum, or, alternatively, free with the centre point fixed, as with a disc in a disc brake assembly. The nodal lines are now *circular* or *radial* (as we showed in the television programme for *Unit 11*).

Extending these ideas still further, we can consider waves in three dimensions. The equation now becomes

$$\frac{\partial^2 \phi}{\partial x^2} + \frac{\partial^2 \phi}{\partial y^2} + \frac{\partial^2 \phi}{\partial z^2} = \frac{1}{c^2}\frac{\partial^2 \phi}{\partial t^2}.$$

The imposition of boundary conditions again gives rise to normal modes, each having a characteristic frequency. The nodes are now *nodal surfaces.*

Typical situations involving the solution of the three-dimensional wave equation include the study of the acoustics of a concert hall, and of electromagnetic radiation in conducting cavities. The equation is also indispensable in the study of atomic structure. The positions of the electrons around the nucleus are governed by certain wave-like properties of the electrons.* Because the electrons are held close to the nucleus by electrostatic attraction, the "waves" have significant amplitude only in the vicinity of the nucleus; they act, therefore, as though subject to boundary conditions. The "boundary" is spherically symmetric about the nucleus, so it becomes more convenient to re-express the wave equation in spherical polar co-ordinates. When this equation is solved, normal nodes are again encountered. This has the effect that the electrons can occupy only certain positions relative to the nucleus. In other words, normal nodes determine the overall configuration of the atom, and hence the interaction of one atom with another. Such interactions determine the chemical properties of the atom. Thus we arrive at the remarkable conclusion that much about the structure of matter can be understood ultimately, at the atomic level, in terms of normal modes—modes that can be analysed by means analogous to Fourier's method.

* These properties are discussed in *Unit S100 29, Quantum Theory* and *Unit S100 30, Quantum Physics and the Atom.*

12.2.7 Summary

In section 12.2 we have shown how the equation of motion

$$\frac{\partial^2 y}{\partial x^2} = \frac{1}{c^2}\frac{\partial^2 y}{\partial t^2}$$

is derived for transverse waves on a string and for longitudinal waves in, for example, air or a rod. The equation is known as the *wave equation in one dimension*. We have shown that functions of the form $f(x - ct)$ and $g(x + ct)$ represent travelling wave solutions of this equation. Since the equation is linear, $\alpha f + \beta g$ is also a solution, where α and β are arbitrary constants. When two harmonic waves of equal amplitude and time period are travelling in opposite directions, a stationary wave is formed; it is represented by the equation

$$y = 2A \cos px \cos pct.$$

There are points (values of x) where the amplitude of this stationary wave is zero at all times; these are called *nodes*. Points where the amplitude is a maximum are called *anti-nodes*.

When *boundary conditions* are imposed, it is found that the frequency of the stationary wave can only take certain values. For example, in the case of a string of length L fixed at both ends, the boundary conditions are $y = 0$ at $x = 0$ and at $x = L$. These conditions limit the permitted frequencies to $nc/2L$, where n is a positive integer and c is the velocity of a travelling wave on the string. The lowest permitted frequency is called the *fundamental*; the others are *overtones*. The solutions of the wave equation that represent these stationary waves are known as *normal modes*.

We pointed out that any periodic vibration of the system can be represented by the sum of normal modes. The amplitudes and phases of the contributing modes are governed by the *initial conditions*. Initial conditions specify the displacement and velocity of each part of the system at time $t = 0$. This brought us to another application of Fourier analysis. The same technique that was used to analyse a periodic forcing function $G(t)$ in terms of sine and cosine components can also be used to analyse the periodic vibration of the system in terms of sines and cosines (i.e. in terms of normal modes). Instead of the harmonic components being functions of time, they are now functions of the distance along the string. This requires a change in the interval over which the integral is taken; $[-\pi, \pi]$ is replaced by $[0, L]$.

12.3 SOLUTIONS TO SELF-ASSESSMENT QUESTIONS

Solution to SAQ 1

Using Equations (3) we find that

$$A_0 = \frac{1}{\pi}\int_{-\pi}^{0} 0\,dt + \frac{1}{\pi}\int_{0}^{\pi} \sin t\,dt = \frac{2}{\pi},$$

and

$$A_n = \frac{1}{\pi}\int_{0}^{\pi} \sin t \cos nt\,dt.$$

Integrating by parts, we obtain

$$\begin{aligned} A_n &= \frac{1}{\pi}\left[\frac{\sin t \sin nt}{n}\right]_0^\pi - \frac{1}{\pi}\int_0^\pi \frac{\cos t \sin nt}{n}\,dt \\ &= 0 + \frac{1}{\pi}\left[\frac{\cos t \cos nt}{n^2}\right]_0^\pi + \frac{1}{\pi}\int_0^\pi \frac{\sin t \cos nt}{n^2}\,dt \\ &= -\frac{1+\cos n\pi}{\pi n^2} + \frac{1}{n^2}A_n. \end{aligned}$$

(This is a rather subtle way of manipulating the integral to get what we want. Instead, we could have used the trigonometric identity

$$2 \sin \alpha \cos \beta = \sin(\alpha + \beta) + \sin(\alpha - \beta)$$

before integrating.)

Thus

$$A_n\left(1 - \frac{1}{n^2}\right) = -\frac{1 + (-1)^n}{\pi n^2},$$

so

$$A_n = -\frac{1 + (-1)^n}{\pi(n^2 - 1)}.$$

It follows that

$$A_n = \begin{cases} 0 & n \text{ odd} \\ \dfrac{-2}{\pi(n^2-1)} & n \text{ even.} \end{cases}$$

Similarly,

$$B_n = \frac{1}{\pi}\int_0^\pi \sin x \sin nx\,dx = \begin{cases} 0 & (n \neq 1) \\ \frac{1}{2} & (n = 1). \end{cases}$$

$$\therefore\quad G(t) = \frac{1}{\pi} + \tfrac{1}{2}\sin t - \frac{2}{\pi}\sum_{n=1}^{\infty}\frac{\cos 2nt}{(4n^2 - 1)}.$$

Solution to SAQ 2

Using Equations (3), we find that

$$A_0 = \frac{1}{\pi}\int_{-\pi}^{\pi} t^2\,dt = \frac{2}{3}\pi^2,$$

and

$$\begin{aligned}
A_n &= \frac{1}{\pi}\int_{-\pi}^{\pi} t^2 \cos nt\, dt \\
&= \frac{1}{\pi}\left[t^2 \frac{\sin nt}{n}\right]_{-\pi}^{\pi} - \frac{2}{\pi}\int_{-\pi}^{\pi} \frac{t}{n} \sin nt\, dt \\
&= 0 - \frac{2}{\pi}\int_{-\pi}^{\pi} \frac{t}{n} \sin nt\, dt \\
&= \frac{2}{\pi}\left[\frac{t \cos nt}{n^2}\right]_{-\pi}^{\pi} - \frac{2}{\pi}\int_{-\pi}^{\pi} \frac{\cos nt}{n^2}\, dt \\
&= \frac{4}{n^2}\cos n\pi - 0 \\
&= \frac{4}{n^2}(-1)^n.
\end{aligned}$$

Similarly,

$$B_n = \frac{1}{\pi}\int_{-\pi}^{\pi} t^2 \sin nt\, dt = 0.$$

$$\therefore \quad G(t) = t^2 = \frac{\pi^2}{3} + 4\sum_{n=1}^{\infty} \frac{(-1)^n \cos nt}{n^2} \qquad t \in [-\pi, \pi].$$

If, in this expansion, we put

$$t = \pi, \quad \cos n\pi = (-1)^n,$$

we obtain

$$\pi^2 = \frac{\pi^2}{3} + 4\sum_{n=1}^{\infty} \frac{1}{n^2}.$$

$$\therefore \quad \sum_{n=1}^{\infty} \frac{1}{n^2} = \frac{\pi^2}{6}.$$

Solution to SAQ 3

(i) odd (ii) even (iii) neither (iv) neither (v) odd
(vi) odd (vii) even (viii) even (ix) even (x) even.

Solution to SAQ 4

From Equations (3), we obtain

$$\begin{aligned}
B_n &= \frac{1}{\pi}\int_{-\pi}^{\pi} f(t) \sin nt\, dt \\
&= \frac{1}{\pi}\int_{-\pi}^{0} f(t) \sin nt\, dt + \frac{1}{\pi}\int_{0}^{\pi} f(t) \sin nt\, dt.
\end{aligned}$$

Putting $t = -x$ in the first integral, we get

$$B_n = \frac{1}{\pi}\int_{\pi}^{0} f(-x) \sin n(-x)\, d(-x) + \frac{1}{\pi}\int_{0}^{\pi} f(t) \sin nt\, dt.$$

Now f is even, so $f(x) = f(-x)$, and hence

$$B_n = \frac{1}{\pi}\int_{\pi}^{0} f(x)(-\sin nx)(-1)\,dx + \frac{1}{\pi}\int_{0}^{\pi} f(t)\sin nt\,dt$$
$$= -\frac{1}{\pi}\int_{0}^{\pi} f(x)\sin nx\,dx + \frac{1}{\pi}\int_{0}^{\pi} f(t)\sin nt\,dt.$$

This expression is zero since the values of the two terms are equal but of opposite sign. Hence we have shown that the coefficient of $\sin nt$ vanishes.

Solution to SAQ 5

The function is odd and therefore we need only the sine terms. From Equations (3), we have

$$B_n = \frac{1}{\pi}\int_{0}^{\pi} \sin nt\,dt + \frac{1}{\pi}\int_{-\pi}^{0} (-\sin nt)\,dt$$
$$= \frac{2}{\pi}\int_{0}^{\pi} \sin nt\,dt \quad \text{(since } \sin nt \text{ is odd)}$$
$$= \frac{2}{\pi}\left[-\frac{\cos nt}{n}\right]_{0}^{\pi}$$
$$= \frac{2}{\pi n}((-1)^{n+1} - (-1))$$
$$= \begin{cases} \dfrac{4}{\pi n} & n \text{ odd} \\ 0 & n \text{ even.} \end{cases}$$

$$\therefore \quad G(t) = \frac{4}{\pi}\left(\sin t + \frac{1}{3}\sin 3t + \frac{1}{5}\sin 5t + \ldots\right).$$

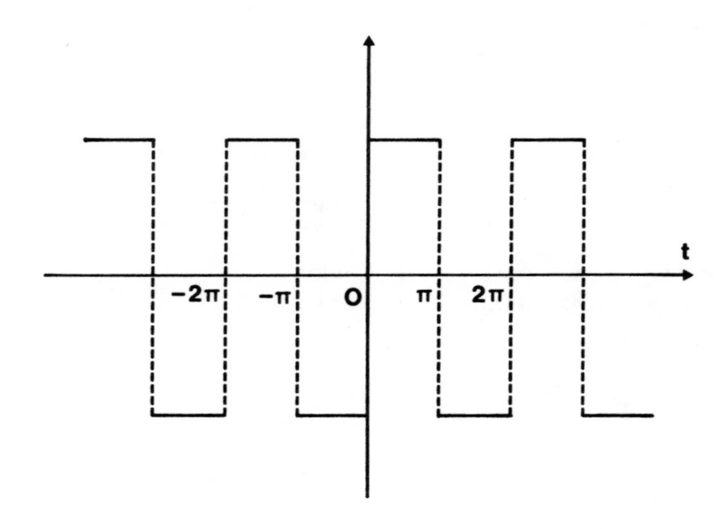

Solution to SAQ 6

The function is even and therefore there are no sine terms. Thus

$$A_0 = \frac{1}{\pi}\left(\int_{0}^{\pi} \sin t\,dt + \int_{-\pi}^{0} (-\sin t)\,dt\right)$$
$$= \frac{2}{\pi}\int_{0}^{\pi} \sin t\,dt = \frac{2}{\pi}\Big[-\cos t\Big]_{0}^{\pi} = \frac{4}{\pi}.$$

Similarly,

$$A_n = \frac{2}{\pi}\int_0^{\pi} \sin t \cos nt \, dt$$

$$= \frac{1}{\pi}\int_0^{\pi} (\sin (n+1)t - \sin (n-1)t) \, dt,$$

where we have used the trigonometric identity

$$\sin \alpha - \sin \beta = 2 \cos \frac{\alpha + \beta}{2} \sin \frac{\alpha - \beta}{2}$$

instead of integrating by parts as in the solution to SAQ 1.

Thus we obtain

$$A_n = \frac{1}{\pi}\left[\frac{\cos (n-1)t}{n-1} - \frac{\cos (n+1)t}{n+1}\right]_0^{\pi} \qquad (n \neq 1)$$

$$= \frac{1}{\pi}\left(\frac{\cos (n-1)\pi}{n-1} - \frac{\cos (n+1)\pi}{n+1} - \frac{1}{n-1} + \frac{1}{n+1}\right)$$

$$= \frac{1}{\pi}\left(\frac{(-1)^{n-1}}{n-1} - \frac{1}{n-1} + \frac{1}{n+1} - \frac{(-1)^{n+1}}{n+1}\right)$$

$$= \frac{1}{\pi}\left(\frac{(-1)^{n-1} - 1}{n-1} + \frac{1 + (-1)^n}{n+1}\right).$$

Also,

$$A_1 = \frac{2}{\pi}\int_0^{\pi} \cos t \sin t \, dt = 0.$$

So we have

$$A_n = \begin{cases} 0 & n \text{ odd} \\ \dfrac{-4}{(n^2 - 1)\pi} & n \text{ even.} \end{cases}$$

Hence

$$G(t) = \frac{2}{\pi} - \left(\frac{4}{3\pi} \cos 2t + \frac{4}{15\pi} \cos 4t + \frac{4}{35\pi} \cos 6t + \cdots\right)$$

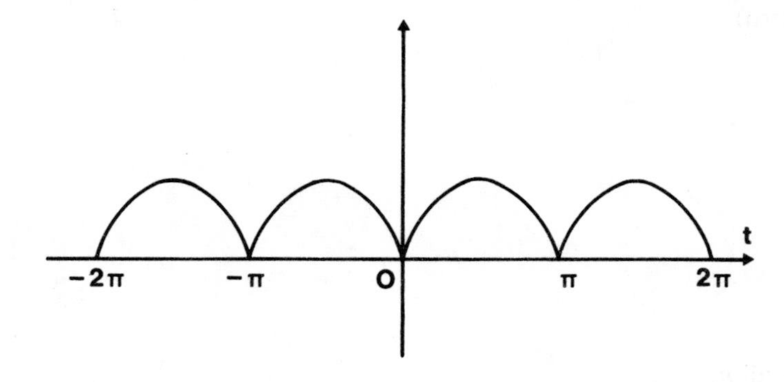

Solution to SAQ 7

There is a slight difference here from the analysis we have so far developed—namely the period of F is $2\pi/\omega$ not 2π, and the analysis is over the interval $[0, 2\pi/\omega]$ instead of $[-\pi, \pi]$. We should expect to make some change in the ranges of integration to allow for this. (See the Additional Notes on p. 38.) The calculation itself is straightforward.

Using Equations (13) of p. S186, we obtain

$$\begin{aligned}
A_0 &= \frac{\omega}{\pi}\int_0^{2\pi/\omega} F(t)\,dt \\
&= \frac{\omega}{\pi}\int_0^{\pi/2\omega} F(t)\,dt + \frac{\omega}{\pi}\int_{\pi/2\omega}^{3\pi/2\omega} F(t)\,dt + \frac{\omega}{\pi}\int_{3\pi/2\omega}^{2\pi/\omega} F(t)\,dt \\
&= 0 + \frac{\omega}{\pi}\int_{\pi/2\omega}^{3\pi/2\omega} F_0\,dt + 0 \\
&= F_0.
\end{aligned}$$

Similarly,

$$\begin{aligned}
A_n &= \frac{\omega}{\pi}\int_{\pi/2\omega}^{3\pi/2\omega} F_0 \cos n\omega t\,dt \\
&= \frac{\omega}{\pi} F_0 \left[\frac{\sin n\omega t}{n\omega}\right]_{\pi/2\omega}^{3\pi/2\omega} \\
&= \frac{F_0}{\pi n}\left(\sin\frac{3n\pi}{2} - \sin\frac{n\pi}{2}\right).
\end{aligned}$$

Now if n is even, $A_n = 0$, since the term in brackets is zero. Suppose n is odd, i.e. $n = 2r + 1$ for some integer r.

Then

$$A_{2r+1} = \frac{F_0}{\pi(2r+1)}\left(\sin\left(3r\pi + \frac{3\pi}{2}\right) - \sin\left(r\pi + \frac{\pi}{2}\right)\right).$$

Expanding the term in brackets, using trigonometric identities, we obtain

$$\begin{aligned}
A_{2r+1} &= \frac{F_0}{\pi(2r+1)} \\
&\qquad \left(\sin 3r\pi \cos\frac{3\pi}{2} + \cos 3r\pi \sin\frac{3\pi}{2} - \sin r\pi \cos\frac{\pi}{2} - \cos r\pi \sin\frac{\pi}{2}\right) \\
&= \frac{F_0}{\pi(2r+1)}(-\cos 3r\pi - \cos r\pi).
\end{aligned}$$

Now we see that if r is even (or zero)

$$A_{2r+1} = \frac{-2F_0}{\pi(2r+1)},$$

whilst if r is odd

$$A_{2r+1} = \frac{2F_0}{\pi(2r+1)}.$$

Integration shows that $B_n = 0$ for all n.

Hence we have

$$F(t) = \frac{1}{2}F_0 + \frac{2F_0}{\pi}\left(-\cos\omega t + \frac{1}{3}\cos 3\omega t - \frac{1}{5}\cos 5\omega t + \cdots\right).$$

From p. S187, Equations (14), (15) and (16), the sustained response of the system is

$$x = \frac{F_0}{2k} + \frac{2F_0}{\pi}\left(\frac{-\cos(\omega t - \alpha_1)}{((k - m\omega^2)^2 + c^2\omega^2)^{1/2}} + \frac{\cos(3\omega t - \alpha_3)}{3((k - 9m\omega^2)^2 + 9c^2\omega^2)^{1/2}} \cdots\right),$$

where

$$\tan \alpha_n = \frac{cn\omega}{k - mn^2\omega^2}.$$

(The solution in the back of S is wrong. In each expression, the factor $\frac{1}{2}$ is missing from the first term.)

Solution to SAQ 8

B

Solution to SAQ 9

We know that any solution may be written as

$$y = \sum_p (A_p \cos pz + B_p \sin pz) \cos(pct + \varepsilon_p).$$

Boundary conditions are

(i) $\dfrac{\partial y}{\partial z} = 0$ at $z = 0$ for all t;

(ii) $\dfrac{\partial y}{\partial z} = 0$ at $z = L$ for all t.

Now (i) gives

$$B_p = 0 \text{ for all } p,$$

and (ii) gives

$$p = \frac{n\pi}{L} \qquad (n = 0, 1, 2, \ldots).$$

Therefore

$$y = \sum_{n=1}^{\infty} D_n \cos\left(\frac{n\pi z}{L}\right) \cos\left(\frac{n\pi ct}{L} + \varepsilon_n\right).$$

D_n and ε_n are undetermined until we know the initial conditions.

Solution to SAQ 10

$$y = \sum_p (A_p \cos px + B_p \sin px) \cos(pct + \varepsilon_p).$$

The boundary conditions are

(i) $y = 0$ at $x = 0$ for all t
(ii) $y = 0$ at $x = L$ for all t.

The initial conditions are

(iii) $y = 0$ at $t = 0$ for all x

(iv) at $t = 0$, $\frac{\partial y}{\partial t} = f(x)$, where

$$f(x) = \begin{cases} 0 & x \in [0, a - \delta] \\ V & x \in [a - \delta, a + \delta] \\ 0 & x \in [a + \delta, L]. \end{cases}$$

Condition (i) is satisfied if $A_p = 0$ for all p.

Condition (ii) is satisfied if $p = \frac{n\pi}{L}$ (where $n = 1, 2, 3, \ldots$).

Condition (iii) is satisfied if $\varepsilon_p = -\frac{\pi}{2}$ for all p.

Thus we have

$$y = \sum_{n=1}^{\infty} B_n \sin \frac{n\pi x}{L} \sin \frac{n\pi ct}{L}.$$

At $t = 0$

$$\frac{\partial y}{\partial t} = f(x) = \sum_{n=1}^{\infty} E_n \sin \frac{n\pi x}{L},$$

where $E_n = \frac{n\pi c B_n}{L}$.

As $\frac{\partial y}{\partial t}$ has this form at $t = 0$, we wish to obtain a similar expansion for $f(x)$—i.e. one consisting of sine terms only. We therefore define $f(x) = -f(-x), x \in [-L, 0]$, so that f is odd, and proceed with the usual Fourier analysis.

Therefore

$$\begin{aligned} E_n &= \frac{2}{L} \int_0^L f(x) \sin \frac{n\pi x}{L}\, dx \\ &= \frac{2}{L} \int_{a-\delta}^{a+\delta} V \sin \frac{n\pi x}{L}\, dx \\ &= \frac{2V}{L} \frac{L}{n\pi} \left[\cos \frac{n\pi}{L}(a - \delta) - \cos \frac{n\pi}{L}(a + \delta) \right] \\ &= \frac{4V}{n\pi} \sin \frac{n\pi a}{L} \sin \frac{n\pi \delta}{L}. \end{aligned}$$

Thus

$$B_n = \frac{4VL}{n^2 \pi^2 c} \sin \frac{n\pi a}{L} \sin \frac{n\pi \delta}{L}$$

and

$$y = \frac{4VL}{\pi^2 c} \sum_{n=1}^{\infty} \frac{1}{n^2} \sin \frac{n\pi a}{L} \sin \frac{n\pi \delta}{L} \sin \frac{n\pi x}{L} \sin \frac{n\pi ct}{L}.$$

If an/L is an integer, the nth term in the series is zero because $\sin (n\pi a/L) = 0$.

Therefore the nth harmonic is missing if an/L is an integer.

Solution to SAQ 11

D.

It may appear that the functions we *use* in the expansion are odd, but, since the function determining the shape is not defined for $x < 0$, we cannot tell.

Solution to SAQ 12

(i) The separation between two adjacent anti-nodes or between two nodes is $\frac{\lambda}{2}$, where λ is the wavelength. The separation between an adjacent anti-node and node is $\frac{\lambda}{4}$.

Pipe *A* has an anti-node at each end so its length is $\frac{\lambda_a}{2}$ in terms of the wavelength λ_a of the fundamental. Pipe *B* has a node at one end and an anti-node at the other, so in terms of its fundamental, the same length is $\frac{\lambda_b}{4}$. Thus the frequency of the fundamental of pipe *A* is twice that of pipe *B*.

(ii) No.

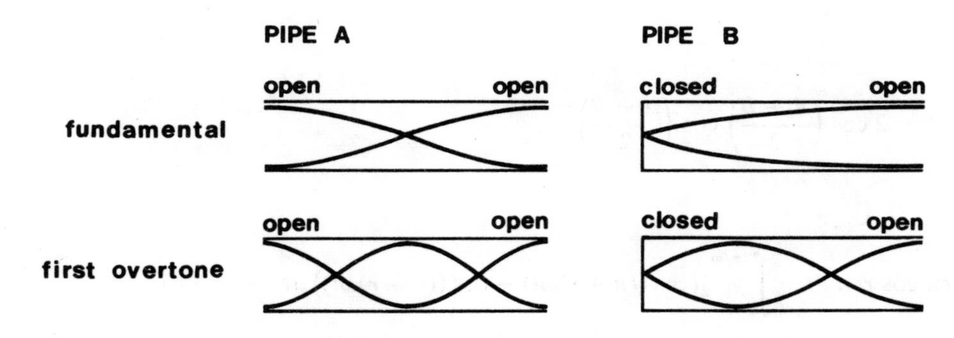

The first overtones in each case are illustrated in the diagram. From this figure you can see that when the first overtone of pipe *A* is sounded, the length of the pipe $= \lambda_a$ (now representing the wavelength of the first overtone). For pipe *B*, the length is $\frac{3\lambda_b}{4}$. Thus the ratio of the two first overtones is $\frac{4}{3}$. This is *not* the same as the ratio of the fundamentals.

(iii) $\lambda = 2 \times$ the length of the pipe.

(iv) No. The ratios of the overtone frequencies to the fundamental frequencies are different in the two cases. Adjusting the length of pipe *B* may cause the fundamental frequency of *B* to coincide with that of *A*, but the overtone frequencies will *not* coincide.

APPENDIX

Additional Notes for Smith and Smith, *Mechanics*

In this appendix we are providing some additional explanatory notes for the parts of the set book we have asked you to read.

If you have found any parts difficult, you should turn to the appropriate page reference of S. These page and line references are on the left-hand side of the pages of this appendix. In some cases we indicate, for example, where proofs can be found in other books, and in other cases we add some further explanation of our own.

Page 186, line 4

This equation simply states that F has constant period $2\pi/\omega$ (ω is, of course, constant). We should also say that F must be "sufficiently well-behaved" before we proceed. In this context, "well-behaved" functions will be piecewise-continuous. This property is too difficult to discuss here, and we shall assume that all functions we need to analyse have this property.

Page 186, line 14

Using the identity

$$\cos A + \cos B = 2\cos\left(\frac{A+B}{2}\right)\cos\left(\frac{A-B}{2}\right),$$

we obtain

$$\begin{aligned}\int_0^{2\pi/\omega} \cos n\omega t \cos r\omega t\, dt &= \int_0^{2\pi/\omega} \tfrac{1}{2}(\cos((n+r)\omega t) + \cos((n-r)\omega t))\, dt \\ &= \frac{1}{2}\left[\frac{\sin((n+r)\omega t)}{(n+r)\omega} + \frac{\sin((n-r)\omega t)}{(n-r)\omega}\right]_0^{2\pi/\omega} \\ &= 0,\end{aligned}$$

since n and r are integers, and $n \neq r$.

Use of similar identities will establish the other equations.

Page 186, line −11

We integrate both sides of Equation (12) from 0 to $2\pi/\omega$. Thus

$$\int_0^{2\pi/\omega} F(t)\, dt = \int_0^{2\pi/\omega} \left(\frac{1}{2}a_0 + \sum_{n=1}^{\infty}(a_n \cos n\omega t + b_n \sin n\omega t)\right) dt.$$

We now assume that the right-hand side can be integrated "term by term". Again this procedure is not always justified, but we shall assume it to be valid for the functions we are interested in. Then we have

$$\int_0^{2\pi/\omega} F(t)\, dt = \frac{\pi}{\omega}a_0 + \sum_{n=1}^{\infty}\left(\int_0^{2\pi/\omega} a_n \cos n\omega t\, dt\right) + \sum_{n=1}^{\infty}\left(\int_0^{2\pi/\omega} b_n \sin n\omega t\, dt\right).$$

By the previous equations, all the terms in the infinite sums are zero. Thus

$$a_0 = \frac{\omega}{\pi}\int_0^{2\pi/\omega} F(t)\, dt.$$

If we now multiply Equation (12) by $\cos n\omega t$ and integrate, the equations on lines 13–16 show that the only non-zero term on the right-hand side is the one involving a_n. Similarly, we can calculate b_n to complete the derivation of Equations (13).

Page 187, line 3

This equation is derived as follows. We wish to solve the equation

$$m\ddot{x}' + c\dot{x}' + kx' = F(t), \tag{i}$$

where $F(t)$ is given by Equation (12). We know that it is sufficient to find a solution for each component of the right-hand side of Equation (12), and then add all these solutions to obtain the solution to (i) above.

The $\frac{1}{2}a_0$ term gives the equation

$$m\ddot{x}' + c\dot{x}' + kx' = \tfrac{1}{2}a_0,$$

which we know has the solution

$$x' = \frac{a_0}{2k}.$$

A term $a_n \cos n\omega t$ gives the equation

$$m\ddot{x}' + c\dot{x}' + kx' = a_n \cos n\omega t. \tag{ii}$$

From pp. S181–182 we see that this equation has the solution

$$x' = \frac{a_n \cos (n\omega t - \phi_n)}{m((\Omega^2 - n^2\omega^2)^2 + 4b^2\Omega^2 n^2\omega^2)^{1/2}}, \tag{iii}$$

where

$$\tan \phi_n = \frac{2b\Omega n\omega}{(\Omega^2 - n^2\omega^2)}$$

and

$$\Omega^2 = \frac{k}{m} \quad \text{and} \quad b = \frac{c}{2\sqrt{mk}}.$$

Rewriting (iii) in terms of the more convenient Q_n, we have

$$x' = a_n Q_n \cos (n\omega t - \phi_n).$$

In the same way, a term $b_n \sin n\omega t$ from Equation (12) leads to the equation

$$m\ddot{x}' + c\dot{x}' + kx' = b_n \sin n\omega t,$$

and the same method of solution yields

$$x' = b_n Q_n \sin (n\omega t - \phi_n)$$

where again,

$$\tan \phi_n = \frac{2b\Omega n\omega}{(\Omega^2 - n^2\omega^2)}.$$

Now adding these solutions for all $n \geqslant 1$ we obtain

$$x' = \frac{a_0}{2k} + \sum_{n=1}^{\infty} Q_n(a_n \cos (n\omega t - \phi_n) + b_n \sin (n\omega t - \phi_n)),$$

as required.

Page 187 line 9

If $b \ll 1$,

$$Q_n \simeq \frac{1}{m(\Omega^2 - n^2\omega^2)},$$

and Q_n becomes larger as ω approaches Ω/n.

Page 188, lines 2–6

The triangular wave is symmetric about π/ω, so we may use the fact that

$$\int_0^{2\pi/\omega} F(t)\,dt = 2\int_0^{\pi/\omega} F(t)\,dt.$$

Glossary

Terms which are defined in this glossary are printed in CAPITALS. Some terms not defined in this list may be found in the *Mathematical Handbook*.

		Page
ANTI-NODE	An ANTI-NODE is a point at which the displacement of a wave is a maximum.	20
AMPLITUDE	The AMPLITUDE of a wave is the maximum value of the displacement.	19
BOUNDARY CONDITIONS	See NORMAL MODE.	
EVEN FUNCTION	An EVEN FUNCTION F has the property that $F(t) = F(-t)$ for all t.	13
FOURIER ANALYSIS	FOURIER ANALYSIS is the process of obtaining an expansion of a PERIODIC FUNCTION G of period 2π in the form $G(t) = \frac{1}{2}A_0 + A_1 \cos t + A_2 \cos 2t + A_3 \cos 3t + \cdots + B_1 \sin t + B_2 \sin 2t + B_3 \sin 3t + \cdots$; such an expansion is called a FOURIER SERIES and the coefficients A_n, B_n are the FOURIER COEFFICIENTS.	8
FOURIER COEFFICIENTS	See FOURIER ANALYSIS.	
FOURIER SERIES	See FOURIER ANALYSIS.	
FUNDAMENTAL MODE	The FUNDAMENTAL MODE is the NORMAL MODE which corresponds to $n = 1$.	22
HARMONIC WAVE	A HARMONIC WAVE is one in which the displacement is given by $y = A \sin p(x \pm ct)$ or $y = B \cos p(x \pm ct)$.	19
INITIAL CONDITIONS	See NORMAL MODE.	
NODE	A NODE is a point at which the displacement of a wave is zero for all time.	20

		Page
NORMAL MODE	One form of the solution of a one-dimensional WAVE EQUATION is $$y = \sum_p (A_p \cos px + B_p \sin px)(\cos pct + C_p \sin pct).$$ If BOUNDARY CONDITIONS are imposed, for example, in the case of a vibrating string, if both ends of the string are fixed, then this equation simplifies further, and only certain values of p will give solutions. In the case of a string fixed at $x = 0$ and at $x = L$, the general solution becomes $$y = \sum_{n=1}^{\infty} B_n \sin\left(\frac{n\pi x}{L}\right) \cos\left(\frac{n\pi ct}{L} + \varepsilon_n\right).$$ Each of the terms in this sum is a NORMAL MODE. If INITIAL CONDITIONS are now supplied, the coefficients B_n and the constants ε_n may be calculated.	22
ODD FUNCTION	An ODD FUNCTION G has the property that $G(-t) = -G(t)$ for all t.	13
OVERTONE	Each NORMAL MODE which corresponds to an integer n ($n > 1$) is called an OVERTONE.	22
PERIOD	See PERIODIC FUNCTION.	
PERIODIC FUNCTION	A PERIODIC FUNCTION G has the property that $G(t + nT) = G(t) \qquad (n \in Z),$ where T is the PERIOD.	7
STATIONARY WAVE	A STATIONARY WAVE is a wave whose profile does not move along the x-axis in time.	20
WAVE EQUATION	The one-dimensional WAVE EQUATION is $$\frac{\partial^2 y}{\partial x^2} = \frac{1}{c^2}\frac{\partial^2 y}{\partial t^2}.$$ c represents the velocity of the wave. This equation is linear.	17

MECHANICS AND APPLIED CALCULUS

1 Some Basic Tools
2 Kinematics
3 Newton's Laws of Motion
4 A Field of Force
5 NO TEXT
6 Rigid Bodies
7 Work and Energy I
8 Work and Energy II
9 Equilibrium and Stability
10 Mechanical Vibrations I
11 Mechanical Vibrations II
12 Fourier Analysis and Normal Modes
13 Rotating Frames of Reference I
14 Rotating Frames of Reference II
15 Variable Mass
16 Orbits
17 NO TEXT